Tamara G. Stryzhak

Frequency Criterion of Stability

Perseval equality:
$$\int_0^\infty y^2(t)dt = \frac{1}{2\pi} \int_{-\infty}^\infty |f(i\omega)|^2 \, d\omega$$

Tamara G. Stryzhak

THE FREQUENCY CRITERION OF STABILITY

ibidem-Verlag

Stuttgart

Bibliografische Information der Deutschen Nationalbibliothek
Die Deutsche Nationalbibliothek verzeichnet diese Publikation in der Deutschen Nationalbibliografie; detaillierte bibliografische Daten sind im Internet über http://dnb.d-nb.de abrufbar.

Bibliographic information published by the Deutsche Nationalbibliothek
Die Deutsche Nationalbibliothek lists this publication in the Deutsche Nationalbibliografie; detailed bibliographic data are available in the Internet at http://dnb.d-nb.de.

∞

Gedruckt auf alterungsbeständigem, säurefreien Papier
Printed on acid-free paper

ISBN-10: 3-8382-0019-5
ISBN-13: 978-3-8382-0019-4

© *ibidem*-Verlag
Stuttgart 2009

Alle Rechte vorbehalten

Das Werk einschließlich aller seiner Teile ist urheberrechtlich geschützt. Jede Verwertung außerhalb der engen Grenzen des Urheberrechtsgesetzes ist ohne Zustimmung des Verlages unzulässig und strafbar. Dies gilt insbesondere für Vervielfältigungen, Übersetzungen, Mikroverfilmungen und elektronische Speicherformen sowie die Einspeicherung und Verarbeitung in elektronischen Systemen.

All rights reserved. No part of this publication may be reproduced, stored in or introduced into a retrieval system, or transmitted, in any form, or by any means (electronic, mechanical, photocopying, recording or otherwise) without the prior written permission of the publisher. Any person who does any unauthorized act in relation to this publication may be liable to criminal prosecution and civil claims for damages.

Printed in Germany

Series: "Modern Mathematics for Engineers"

Lectures for the trainee-students of IAESTE

Educational publication

Lectures
Professor NTUU "KPI" Tamara Stryzhak

Translator
**Faculty member NTUU "KPI"
Nataliya Sarycheva**

English language editors
**IAESTE trainee students:
Betina Carina Siebert (Germany)
Laura Lambertz (Germany)
Mirjana Vukelja (Swizerland)**

Series:
"Modern Mathematics for Engineers"

Frequency Criterion of Stability

**Professor NTUU "KPI"
Tamara Stryzhak**

Lectures

There are different criteria for the stability of the solution of linear differential equations, which are based on the characteristic equation analysis. In this work the reader will find principally new ways of frequency criterion for stability derivation, which are based on Perceval formula and resolution of operator equations in Banach space. At first we consider simple examples and then a more general theory will be stated.

All remarks and recommendations are greatly appreciated and will be taken into considerations in subsequent editions.

Phone: +380 44 2418648
Fax: +380 44 2418648, +380 44 2417620
E-mail: stri@aer.ntu-kpi.kiev.ua
Web-site: www.iaeste.org.ua

© Stryzhak T. G., 2009

Contents

§1 Perceval Equality — 12

§2 Dependence of a characteristic equation on a parameter — 17

§3 L_2-stability of the solutions of differential equations — 24

§4 Frequency Stability Criterion by V.M.Popov — 34

§5 Stability and Instability Criteria for the Solutions of a System of Differential-difference Equations — 40

§6 Criteria for Keeping Exponential Dichotomy — 49

§7 Method of the Transfer Function — 58

§8 The Transfer Function of the Non-stationary System of Differential Equations — 65

§9 Generalization of the transfer function notion — 72

§10 Successive approximation method — 75

Appendix: Tasks for solving 79
Literature 81
Project "Modern mathematics for Engineers" 83

Frequency Criterion of Stability

There are different criteria for the linear differential equation solution stability, which are based on the characteristic equation analysis [1]. In this work the reader will find principally new ways of frequency criterion for stability derivation, which are based on Perceval formula and resolution of operator equations in Banach space [2]. At first we shall consider simple examples and then a more general theory will be stated.

§1 Perceval Equality

Let $y(t)$ be a continuous, integrated on $[0,+\infty)$, function which leads to the improper integral

$$\int_0^\infty y^2(t)\,dt < \infty. \tag{1}$$

We shall introduce the linear Banach space L_2 with the function $y(t)$, for which the following equality (1) is fulfilled with the condition

$$\|y(t)\| = \left(\int_0^\infty y^2(t)\,dt\right)^{1/2}. \tag{2}$$

Let $f(p)$ be a display of the function $y(t)$ according to Laplace

$$f(p) = \int_0^\infty e^{-pt} y(t)\,dt.$$

Here the Perceval equality will be true

$$\int_0^\infty y^2(t)\,dt = \frac{1}{2\pi}\int_{-\infty}^\infty |f(i\omega)|^2\,d\omega. \tag{3}$$

Example. We shall demonstrate with the following example the validity of the Perceval equation.
Let
$$y(t) = e^{-\alpha t} \quad (\alpha > 0),$$
then
$$\int_0^\infty y^2(t)\,dt = \int_0^\infty e^{-2\alpha t}\,dt = \frac{1}{2\alpha}.$$

We shall find the function $y(t)$ by Laplace
$$f(p) = \int_0^\infty e^{-pt} \cdot e^{-\alpha t}\,dt = \frac{1}{p+\alpha}$$
and calculate the integral
$$\frac{1}{2\pi}\int_{-\infty}^\infty |f(i\omega)|^2\,d\omega = \frac{1}{2\pi}\int_{-\infty}^\infty \frac{d\omega}{|i\omega+\alpha|^2} = \frac{1}{2\pi}\int_{-\infty}^\infty \frac{d\omega}{\alpha^2+\omega^2} = \frac{1}{2\alpha}.$$

It proves the validity of the following equation
$$\int_0^\infty y^2(t)\,dt = \frac{1}{2\pi}\int_{-\infty}^\infty |f(i\omega)|^2\,d\omega.$$

The inequality (1) will be written as follows $y(t) \in L_2$.

If
$|z(t)| \leq m \ (t \geq 0), \ y(t) \in L_2$, then $z(t)y(t) \in L_2$

and
$$\|z(t)y(t)\| \le m\|y(t)\|.$$
Let $g(y)$ be a non-linear function, so $|g(y)| \le m|y|$.

If $y(t) \in L_2$, then $g(y(t)) \in L_2$ and
$$\|g(y(t))\| \le m\|y(t)\|.$$

Let $y(t)$ be a vector with projections $y_1(t),...,y_n(t)$ so $y_k(t) \in L_2$ $(k=1,...,n)$.

We shall introduce the Euclidean norm for the vector $Y(t)$ according to the formula

$$|Y(t)|^2 = \sum_{k=1}^{n} |y_k^2(t)|. \qquad (4)$$

We shall introduce the display of the vector $Y(t)$ by Laplace

$$F(p) = \int_0^\infty e^{-pt} Y(t) dt.$$

The Perceval equality will look like

$$\int_0^\infty |Y(t)|^2 dt = \frac{1}{2\pi} \int_{-\infty}^{\infty} |F(i\omega)|^2 d\omega. \qquad (5)$$

We shall introduce the norm for the vector $Y(t)$ in the space L_2

$$\|Y(t)\| = \left(\int_0^\infty |Y(t)|^2 dt\right)^{1/2}. \tag{6}$$

Let the two functions $y(t)$, $x(t) \in L_2$ be linked by the relation

$$L(d)y(t) = L_1(d)x(t), \tag{7}$$

where $L(d)$, $L_1(d)$ are differential operators

$$L(d) = \sum_{k=0}^n a_k d^k, \quad L_1(d) = \sum_{k=0}^n b_k d^k \quad \left(d \equiv \frac{d}{dt}\right).$$

For displaying functions $y(t)$, $x(t)$ by Laplace

$$f(p) = \int_0^\infty e^{-pt} y(t) dt, \quad g(p) = \int_0^\infty e^{-pt} x(t) dt$$

with zero initial values at $t = 0$:

$$y(0) = 0, \ y'(0) = 0, \ldots, y^{(n-1)}(0) = 0;$$
$$x(0) = 0, \ x'(0) = 0, \ldots, x^{(n-1)}(0) = 0$$

we shall receive the equation

$$L(p)f(p) = L_1(p)g(p). \tag{8}$$

Let us assume $\|x(t)\|$ and find an estimation for $\|y(t)\|$. Out of Perceval equality we shall receive the following estimation

16 Frequency Criterion of Stability

$$\|y(t)\|^2 = \int_0^\infty y^2(t)\,dt = \frac{1}{2\pi}\int_{-i\infty}^{i\infty}|f(i\omega)|^2\,d\omega = \frac{1}{2\pi}\int_{-i\infty}^{i\infty}\left|\frac{L_1(i\omega)}{L(i\omega)}\right|^2|g(i\omega)|^2\,d\omega \leq$$

$$\leq \max_\omega\left|\frac{L_1(i\omega)}{L(i\omega)}\right|^2 \cdot \frac{1}{2\pi}\int_{-i\infty}^{i\infty}|g(i\omega)|^2\,d\omega = \max_\omega\left|\frac{L_1(i\omega)}{L(i\omega)}\right|^2\|x(t)\|^2$$

out of which we shall find the inequality

$$\|y(t)\| \leq \max_\omega\left|\frac{L_1(i\omega)}{L(i\omega)}\right|\cdot\|x(t)\| \qquad (9)$$

§2 Dependence of a characteristic equation on the parameter

We shall consider a differential-difference equation

$$L(d,\varepsilon)y(t)=0, \quad \left(d \equiv \frac{d}{dt}\right), \tag{10}$$

where $L(d,\varepsilon)$ is a differential operator. Let us suppose that the characteristic equation

$$L(p,\varepsilon)=0$$

at $\varepsilon = 0$ has all roots with a negative real part. According to this condition the equation solutions (10) are asymptotically stable. We shall give a criterion which proves that the equation roots $L(p,\varepsilon)=0$ at the changed parameter ε remain in the half-plane $\operatorname{Re} p < 0$.

We shall consider the linear differential equation with constant coefficients

$$a_n y^{(n)} + a_{n-1} y^{(n-1)} + a_{n-2} y^{(n-2)} + \ldots + a_0 y = 0, \quad a_n \neq 0 \tag{11}$$

We shall define the polynomial

$$L(p) = \sum_{k=0}^{n} a_k d^k.$$

If all roots of the equation $L(p)=0$ lie in the left half-plane $\operatorname{Re} p < 0$, then solutions of the equation (11) are asymptotically stable.

We shall consider the following disturbed differential equation

$$a_n y^{(n)} + a_{n-1} y^{(n-1)} + a_{n-2} y^{(n-2)} + \ldots + a_0 y =$$
$$= \varepsilon \left(b_n y^{(n)} + b_{n-1} y^{(n-1)} + b_{n-2} y^{(n-2)} + \ldots + b_0 y \right), \quad (12)$$

where ε is a parameter complex, in a general case.

This equation can be written in an operational form

$$L(d) y(t) = \varepsilon L_1(d) y(t).$$

$$L(d) = \sum_{k=0}^{n} a_k d^k, \quad L_1(d) = \sum_{k=0}^{n} b_k d^k.$$

We shall prove the theorem which is the frequency criterion for stability.

Theorem. Let the zero solution to the differential equation (11) be asymptotically stable.

If the following equality is fulfilled

$$|\varepsilon| \cdot \sup_{\omega} \frac{|L_1(i\omega)|}{|L(i\omega)|} < 1, \qquad (13)$$

the zero solution of the differential equation (12) is also asymptotically stable.

The asymptotical behavior of the linear differential equation (12) is defined by the location of the equation roots

$$L(p) = \varepsilon L_1(p) \qquad (14)$$

At $\varepsilon = 0$ all roots of the equation (14) are located in the left half-plane $\operatorname{Re} p < 0$. At the changed ε the equation roots (14) continuously change their location on the plane. If at some value ε the equation (14) has a pure imaginary root $p = i\omega$, then we shall receive the following equality

$$\frac{|\varepsilon| \cdot |L_1(i\omega)|}{|L(i\omega)|} = 1.$$

According to the inequality (13) it is impossible. So if the condition (13) is fulfilled, all roots of the equation (14) stay in the left half-plane and solutions to the equation (12) remain asymptotically stable.

Example. We shall find conditions of the asymptotic stability of the solutions of the linear differential equation

$$y'' + 2vy' + v^2 y + \alpha y' + \beta y'' = 0, \quad (v > 0) \quad (15)$$

The stability condition (13) will assume the following view

$$\sup_{\omega} \left| \frac{\alpha i\omega + \beta (i\omega)^2}{(i\omega)^2 + 2vi\omega + v^2} \right| < 1.$$

This inequality will be fulfilled if the following inequality is also fulfilled

$$\sup_{\omega} \left| \frac{i\omega}{i\omega + v} \right| \cdot \left(\sup_{\omega} \frac{|\alpha|}{|i\omega + v|} + \sup_{\omega} \left| \frac{i\omega \beta}{i\omega + v} \right| \right) < 1,$$

which will assume a simple view

$$|\alpha| v^{-1} + |\beta| < 1. \quad (16)$$

If the condition (16) is fulfilled the zero solution of the equation (15) will be asymptotically stable.

It is apparent that the stability criterion (13) can be generalized for differential-difference equations without changing the proof.

Theorem. Let the zero solution of the differential-difference equation

$$L(d) y(t) = 0$$

be asymptotically stable. If the following inequality is fulfilled

$$|\varepsilon| \sup_{\omega} \frac{|L_1(i\omega)|}{|L(i\omega)|} < 1, \qquad (17)$$

then the zero solution of the differential-difference equation

$$L(d)y(t) + \varepsilon L_1(d)y(t) = 0$$

will also be asymptotically stable.

Example. We shall find a sufficient condition of the asymptotical stability for the equation solution

$$y''(t) + 2y'(t) + y(t) + \varepsilon y(t-\tau) = 0, \quad (\tau \geq 0).$$

This equation can be recorded in the operational form of

$$(d^2 + 2d + 1) \cdot y(t) + \varepsilon e^{-d\tau} = 0, \quad d \equiv \frac{d}{d\tau}.$$

Out of the inequality (17)

$$|\varepsilon| \sup_{\omega} \frac{|e^{-i\omega\tau}|}{|(i\omega)^2 + 2i\omega + 1|} < 1$$

we shall find the sufficient condition of the asymptotical stability: $|\varepsilon| < 1$. Here ε is a complex parameter.

Example. We shall find a condition of an asymptotic stability of the solution of the difference equation

$$y(t) = \frac{1}{2} y(t-\tau_1) + \varepsilon\, y(t-\tau_2), \quad \tau_1 \geq 0,\ \tau_2 \geq 0.$$

This equation can be written in an operator form

$$y(t) = \frac{1}{2} e^{-d\tau_1} y(t) + \varepsilon e^{-d\tau_2} y(t).$$

The frequency criterion for stability (13) will assume the following view

$$|\varepsilon| \sup_{\omega} \frac{\left|e^{-i\omega\tau_2}\right|}{\left|1 - \frac{1}{2} e^{-i\omega\tau_1}\right|} < 1,$$

out of which we can find the condition of asymptotic stability $|\varepsilon| < 0,5$.

As the condition (17) means that none of the equation roots (14) cross the line $\operatorname{Re} p = 0$, then the following theorem of instability will be true.

Theorem. Let the solutions of the differential-difference equation

$$L(d) y(t) = 0 \quad \left(d \equiv \frac{d}{dt}\right)$$

be unstable. If the following condition is fulfilled

$$|\varepsilon| \sup_{\omega} \frac{|L_1(i\omega)|}{|L(i\omega)|} < 1,$$

then the zero solution of the differential-difference equation

$$L(d)y(t)+\varepsilon L_1(d)y(t)=0$$

is also unstable.

Example. The zero solution of the differential equation

$$y''(t)-y(t)=0$$

is unstable. We shall consider the following differential equation

$$y''(t)-y(t)+\varepsilon y'(t-\tau)=0 \quad (\tau>0).$$

The condition (17) will assume the following view

$$\sup_{\omega}\frac{|\varepsilon|\,|i\omega e^{-i\omega\tau}|}{|\omega^2+1|}<1$$

and is fulfilled at $|\varepsilon|<2$. Consequently, at $|\varepsilon|<2$ the equation solutions with the argument delay are unstable.

§3 L₂ – stability of the differential equation solutions

We shall consider the following linear differential equation with variable coefficients

$$y^{(n)}(t) + \sum_{k=0}^{n-1} a_k(t) y^{(k)}(t) = 0. \qquad (18)$$

This equation can be written in the form $L(t,d)\, y(t) = 0$, where $L(t,d)$ is a differential operator with variable coefficients

$$L(t,d) = d^n + \sum_{k=0}^{n-1} a_k(t)\, d^k.$$

Definition. The linear differential-difference equation (18) is called L_2-stable, if any solution $y(t)$ of the equation (18) belongs to the space L_2, i.e.

$$\int_0^\infty |y(t)|^2\, dt < \infty.$$

If the equation (18) has constant coefficients then it is apparent that L_2-stability is equivalent to asymptotic stability of the equation solution (18).

We shall produce the apparent result without a proof.

Theorem. If the solution coefficients (18) are limited at $t \geq 0$, then out of L_2-stability of the equation solutions (18) comes out asymptotic stability of the equation solutions (18).

Let the differential-difference equation solutions

$$\sum_{k=0}^{n} a_k y^{(k)}(t) = 0$$

be asymptotically stable. We shall consider a disturbed differential equation with variable coefficients

$$\sum_{k=0}^{n} a_k y^{(k)}(t) = \varepsilon x(t), \quad x(t) = b(t) \cdot y(t), \quad (19)$$

where $b(t)$ is a continuous limited at $t \geq 0$ function

$$\left| b(t) \right| \leq b = const.$$

To research the equation (19) we shall apply Laplace transformation. We assume

$$f(p) = \int_0^\infty e^{-pt} y(t) dt, \quad g(p) = \int_0^\infty e^{-pt} x(t) dt$$

and come to the equation for the image $f(p)$

$$L(p) \, t(p) = z(p) + \varepsilon g(p), \quad L(p) \equiv \sum_{k=0}^n a_k p^k,$$

where $z(p)$ is a polynomial of $n-1$ degree, which depends on the initial values $y(0), y'(0), ..., y^{(n-1)}(0)$. Using the definition

$$G(t) = \frac{1}{2\pi i} \int_{\sigma - i\infty}^{\sigma + i\infty} L^{-1}(p) \, e^{pt} dp \quad (\sigma \geq 0), \quad (20)$$

we can find the equation solution (19) as

$$y(t) = y_0(t) + \varepsilon \int_0^t G(t-\tau) \, b(\tau) y(\tau) d\tau. \quad (21)$$

We received an integral equation instead of a differential equation (19). We shall solve the equation (21) by means of subsequent approximations. Here $y_0(t)$ is a solution of a homogeneous equation (19) at $x(t) \equiv 0$. It is apparent that $y_0(t) \in L_2$.

We shall introduce an integral operator A, defined by the following expression

$$Ay(t) \equiv \int_0^t G(t-\tau) \; b(\tau) \; y(\tau) d\tau. \qquad (22)$$

Let $y(t) \in L_2$. We shall find the norm of the operator A. We shall find an image of a function $Ay(t)$ by Laplace

$$\int_0^\infty e^{-pt} Ay(t) dt = \int_0^\infty e^{-pt} \int_0^t G(t-\tau) \; x(\tau) d\tau = L^{-1}(p) g(p)$$

With the help of Perceval formula we find

$$\|Ay(t)\|^2 = \frac{1}{2\pi} \int_{-i\infty}^{i\infty} |L^{-1}(i\omega) g(i\omega)|^2 \, d\omega \le \sup_\omega |L^{-1}(i\omega)|^2 \cdot$$

$$\cdot \frac{1}{2\pi} \int_{-i\infty}^{i\infty} |g(\varepsilon\omega)|^2 \, d\omega = \sup_\omega |L^{-1}(i\omega)|^2 \cdot \|x(t)\|^2.$$

We find estimation for the function $x(t)$

$$\|x(t)\|^2 = \|b(t) y(t)\|^2 = \int_0^\infty |b(t)|^2 \cdot |y(t)|^2 \, dt \le b^2 \|y(t)\|^2.$$

$$(23)$$

Out of the inequality

$$\|Ay(t)\|^2 \le \sup_{\omega}|L^{-1}(i\omega)|^2 b^2 \|y(t)\|^2$$

we find estimation for the norm of the operator A in the space L_2

$$\|A\| \le b \sup_{\omega}|L^{-1}(i\omega)|. \qquad (24)$$

The integral equation (21) can be written as the following operational form

$$y(t) = y_0(t) + \varepsilon\, Ay(t) \qquad (25)$$

And according to the Banah theorem of the operational equations solution, the equation (25) can be solved with the help of the subsequent approximation method.

$$y_{n+1}(t) = y_0(t) + \varepsilon\, Ay_n(t), \quad (n = 0,1,2,...).$$

At $|\varepsilon|\cdot\|A\| < 1$ the subsequent approximations $y_n(t)$ converge in L_2 and there is a solution to the equation (25)

$$y(t) = \lim_{n\to\infty} y_n(t), \quad y(t) \in L_2.$$

Consequently, if the following condition is fulfilled

$$|\varepsilon|\cdot b \cdot \sup_{\omega}|L^{-1}(i\omega)| < 1 \qquad (26)$$

The zero solution of the equation (19) will be L_2-stable.

Notice. We shall receive the same result if the right part of the equation (19) satisfies the following condition

$$x(t) = \psi(t, y(t)), \quad |\psi(t,y)| \le b|y|, \quad (t \ge 0),$$

where $\psi(t,y)$ is a random function which offers the equation solution

$$\sum_{k=0}^{n} a_k y^{(k)}(t) = \varepsilon \psi(t, y(t)).$$

Apparently it is sufficient that the functionO $\psi(t,y)$ was continuous at $t \ge 0$, $|y| < \infty$.

In the same way it is proved:

Theorem. If a zero solution of the differential-difference equation

$$L(d) y(t) = 0$$

is asymptotically stable, then the solution of the non-linear differential-difference equation

$$L(d) y(t) + \varepsilon L_1(d) \psi(t, y(t)) = 0,$$

where the function $\psi(t,y)$ satisfies the condition

$$|\psi(t,y)| \leq b|y| \quad (t \geq 0),$$

will be L_2-stable if the following inequality is fulfilled

$$|\varepsilon| \, b \sup_{\omega} \left| \frac{L_1(i\omega)}{L(i\omega)} \right| < 1. \qquad (27)$$

This theorem is called **frequency criterion for stability**.

Example: We shall find the condition for stability of the linear differential equation solution

$$y'' + 2vy' + y + 2\varepsilon \, y \cos t \cdot y = 0 \quad (v > 0).$$

We have the following equality

$$L(d) = d^2 + 2vd + 1,$$

$$\sup_{\omega} \left| \frac{1}{(i\omega)^2 + 2vi\omega + 1} \right| = \frac{1}{2v\sqrt{1-v^2}}.$$

Out of the inequality (27) we shall find at $0 < v^2 \leq 0{,}5$ the stability condition: $|\varepsilon| < \sqrt{v^2(1-v)^2}$. At $v^2 \geq 0{,}5$ the stability condition will assume the view of the following inequality

$$|\varepsilon| < \frac{1}{2}.$$

Example. We shall research stability of the solution of the differential-difference equation

$$y''(t) + 2vy'(t) + v^2 y(t) + \varepsilon \cos t \cdot y(t-\tau) = 0$$

$$(\tau \geq 0,\ v > 0).$$

We shall find the following operator

$$L(d) = d^2 + 2vd + v^2, \quad L_1(d) = e^{-d\tau}.$$

Out of the condition (27) we shall find the following stability condition

$$|\varepsilon| \cdot \sup_{\omega} \left| \frac{e^{-i\omega\tau}}{(i\omega)^2 + 2v(i\omega) + v^2} \right| < 1.$$

Consequently, according to the condition $|\varepsilon| < v^2$ the solution of the initial differential equation is stable.

Example. We shall find the stability condition of the solution of the differential equation with the lag $\tau = 0,5$:

$$y'(t) - y(t-\tau) = \varepsilon \psi(y(t)), \quad |\psi(y)| \leq b|y|.$$

The stability condition (26) will assume the view of the following inequality

$$|\varepsilon| \cdot b \sup_{\omega} \left| \frac{1}{i\omega - e^{-i\omega\tau}} \right| < 1,$$

out of which we shall find the condition $|\varepsilon|b<1$.

Example. We shall research the stability of solutions of the differential equation
$$y'' + 2vy' + v^2 y + \frac{d}{dt} f(t, y, y') = 0, \quad v > 0,$$
where the function $f(t, y, z)$ satisfies the following condition
$$|f(t, y, z)| \leq \alpha |y| + \beta |z|, \quad (\alpha > 0, \beta > 0).$$

We shall write the equation in the operational form
$$(d+v)^2 y + df(t, y, dy) = 0, \quad d \equiv \frac{d}{dt}.$$

Out of the stability criterion (27) we shall receive the sufficient condition of the solution stability
$$\sup_\omega \left| \frac{i\omega}{i\omega + v} \right| \cdot \left(\sup_\omega \frac{\alpha}{|i\omega + v|} + \sup_\omega \left| \frac{i\omega \beta}{i\omega + v} \right| \right) < 1,$$
which will assume a simple view
$$\alpha v^{-1} + \beta < 1.$$

Example. The stability condition of the zero solution of the differential equation
$$y'' + 2vy' + v^2 y + f(t, y, y') = 0, \quad (v > 0),$$
if the following condition is performed

$$|f(t,y,z)| \leq \alpha |y| + \beta |z|,$$

will look like the following inequality

$$\alpha v^{-1} + \beta < 1.$$

§4 Frequency Stability Criterion by V.M. Popov

We shall consider the system of linear differential equations

$$\frac{dY(t)}{dt} = AY(t) + Bf(C*Y(t)), \quad 0 \leq \frac{f(y)}{y} \leq b$$

where A is a matrix $m \times m$, B, C are vectors of dimension m. This system can be reduced to one particular differential equation

$$L(d)y(t) + f(y(t)) = 0, \quad d \equiv \frac{d}{dt} \quad (28)$$

where $L(d)$ is a fractional-rational function from $d \equiv \frac{d}{dt}$.

In works [3, 4] with the help of Lyapunov functions it was proved that the zero solution of equation (28) is stable if the following inequality is performed

$$\operatorname{Re}\left((1 + qi\omega)L^{-1}(i\omega)\right) + b^{-1} > 0 \quad (29)$$

at some value of the actual parameter q and all values $\omega \notin (-\infty, \infty)$.

Condition (29) can be written in the form
$$b^{-1} > \sup_{\omega}\{-\operatorname{Re}(1+qi\omega)L^{-1}(i\omega)\}. \qquad (30)$$

This condition is sufficient for stability of solution of the equation (28). We shall suppose positiveness of the right part of t inequality (30). The biggest value of parameter b can be found out of the following inequality

$$0 < b < \left(\min_{q}\left\{\sup_{\omega}\{-\operatorname{Re}(1+qi\omega)L^{-1}(i\omega)\}\right\}\right)^{-1} \qquad (31)$$

The inequality (31) can be used at any value of the parameter q.

The condition (29) is called a frequency criterion by V.M. Popov.

We shall consider a more complex differential-difference equation

$$L(d)y(t)+f(t,y(t))=0, \qquad (32)$$

where де $L(d)$ is a differential-difference operator and thenon-linear function $f(t,y)$ satisfies the following condition

$$0 \le f(t,y)y^{-1} \le b \quad (b>0). \qquad (33)$$

In order to use the stability condition (27), we shall write the equation (32) in the form of

$$(L(d)+0,5b)y(t)+(f(t,y(t))-0,5by(t))=0.$$

The following inequality will be performed

$$|f(t,y)-0,5by| \le 0,5b|y|.$$

The stability condition (27) can be reduced to the following inequality

$$\sup_{\omega} \left| \frac{0,5b}{L(i\omega)+0,5b} \right| < 1. \qquad (34)$$

This inequality can be written in the following form

$$\sup_{\omega} \left| \frac{z}{1+z} \right| < 1, \quad z \equiv \frac{0,5b}{L(i\omega)}. \qquad (35)$$

The inequality $\left|\frac{z}{1+z}\right|<1$ or $|z-0|<|z-(-1)|$ has the solution $\operatorname{Re} z > -0,5$. Consequently the inequality (34) can be written in the following form

$$\operatorname{Re} L^{-1}(i\omega) + \frac{1}{b} > 0, \quad b > 0. \qquad (36)$$

The frequency criterion (36) is received as a particular case of frequency stability (29) at $q=0$. The stability criterion (36) is applied for the differential-difference equations (32), which are more complex than the differential equation (28). The stability criterion (36) can be written in the form of the following inequality

$$0 < b < \left(\sup_{\omega} \left(-\operatorname{Re} L^{-1}(i\omega) \right) \right)^{-1}. \qquad (37)$$

The difference of the stability criteria will be shown in the following examples.

Example. It is known that a linear differential equation with constant coefficients

$$y''(t) + ay'(t) + cy(t) = 0$$

is asymptotically stable only at $a > 0$, $c > 0$.

We shall research stability of solutions of the following differential equation

$$y''(t) + y'(t) + b(y(t) - y'(t)) = 0. \qquad (38)$$

Solutions of the equation (38) are unstable at $b < 0$ or $b > 1$, as one of the coefficients of the equation (38) is negative.

The equation (38) can be written in the following form

$$(d^2 + d) \, y(t) + (1-d) f(y) = 0,$$

$$\frac{f(y)}{y} = b, \quad d \equiv \frac{d}{dt}. \qquad (39)$$

38 FREQUENCY CRITERION OF STABILITY

We shall apply the stability criterion (29) for theequation (39). We have the following expression

$$L(i\omega) = \frac{(i\omega)^2 + i\omega}{1 - i\omega},$$

$$\operatorname{Re}(1 + qi\omega) L^{-1}(i\omega) = \frac{(q-2) - \omega^2 q}{1 + \omega^2}.$$

We shall find the following values

$$\sup_{\omega}\{-\operatorname{Re}(1 + qi\omega) L^{-1}(i\omega)\} = \max\{2 - q, q\},$$

$$\min_{q}\{\max\{2 - q, q\}\} = 1 \text{ при } q = 1.$$

Consequently, the frequency stability criterion by V.M. Popov gives the correct answer $0 < b < 1$.

The stability criterion (37) gives the following condition

$$0 < b < \left(\sup_{\omega} \frac{2}{1 + \omega^2}\right)^{-1} = \frac{1}{2}. \qquad (40)$$

Example. We shall find the condition of stability for the solution of the differential equation

$$(d^2 + d) y(t) + b(1 - d) \sin^2 t \cdot y(t) = 0. \qquad (41)$$

The stability criterion by V.M. Popov cannot be applied for the research of the stability of the equation solution (41), and the stability criterion (36) gives the following sufficient stability condition (40)
$$0 < b < 0,5.$$

§5 Stability and Instability Criteria of the Solutions of the System of Differential-difference Equations.

Let $Y(t)$ be a n-dimensional vector. We shall use the following norm

$$|Y(t)| = \left(\sum_{k=1}^{n} |y_k(t)|^2 \right)^{1/2}, \qquad (42)$$

where $y_k(t)$ $(k=1,...,n)$ are projections of the vector $Y(t)$. We shall suppose that $y_k(t) \in L_2$, i.e. the following inequalities are performed

$$\int_{-\infty}^{\infty} y_k^2(t) dt < \infty, \qquad (k=1,...,n).$$

We shall define with the help of $\|Y(t)\|$ the norm of the vector $Y(t)$ in L_2

$$\|Y(t)\|^2 = \int_{-\infty}^{\infty} \sum_{k=1}^{n} |y_k(t)|^2 dt. \qquad (43)$$

With the help of $|B(t)|$ we shall denote the norm of matrix $B(t)$ in the Euclidean space (42).

We shall consider a simplified case of the system of the differential-difference equations

$$L(d)Y(t)+\varepsilon B(t)Y(t)=X(t)$$
$$|B(t)|\leq b,\quad X(t)\in L_2. \tag{44}$$

We shall suppose that at $\varepsilon = 0$ for the solution of a homogeneous system of the equations $L(d)Y(t)=0$ exponential dichotomy of the solutions takes place, i.e.

$$\det L(i\omega)\neq 0,\quad \omega\in(-\infty,\infty).$$

The system of equations for a particular solution of the equation system (44) will assume the following view

$$Y(t)=X_1(t)-\varepsilon\int_{-\infty}^{\infty}G(t-\tau)\cdot B(\tau)\cdot Y(\tau)d\tau, \tag{45}$$

where it is defined as follows

$$X_1(t)\in L_2,\quad G(t)=\frac{1}{2\pi}\int_{-i\omega}^{+i\omega}L^{-1}(p)e^{pt}dp.$$

We shall use the two-sided transformation by Laplace

$$F(p)=\int_{-\infty}^{\infty}Y(t)e^{-pt}dt,\quad Y(t)\in L_2.$$

The Perceval equation is true

$$\int_{-\infty}^{\infty} |Y(t)|^2 \, dt = \frac{1}{2\pi} \int_{-\infty}^{\infty} |F(i\omega)|^2 \, d\omega. \quad (46)$$

We shall solve the equation system (45) with the help of sequential approximations. We shall define the integral operator in the space L_2 as A

$$AY(t) = \int_{-\infty}^{\infty} G(t-\tau) B(\tau) Y(\tau) \, d\tau. \quad (47)$$

The condition of convergence of the method of sequential approximations will look like the following inequality

$$|\varepsilon| \cdot \|A\| < 1.$$

We shall estimate the norm of the operator A (47) in L_2. Let $Y(t) \in L_2$, i.e. the following inequality is performed

$$\|Y(t)\|^2 = \int_{-\infty}^{\infty} |Y(t)|^2 \, dt \le c^2.$$

Then for the vector-function

$$Y_1(t) = B(t) \; Y(t)$$

we have the estimation

$$\|B(t) \; Y(t)\|^2 = \int_{-\infty}^{\infty} |B(t)|^2 \cdot |Y(t)|^2 \, dt \le b^2 \|Y(t)\|^2.$$

Thus, we shall receive a simple relation

$$\|B(t)\ Y(t)\| \le b\ \|Y(t)\|.$$

We shall introduce the Laplace two-sided transformation for $Y_1(t)$

$$F_1(p) = \int_{-\infty}^{\infty} e^{-pt} Y_1(t)\,dt.$$

According to the convolution theorem we shall find the following equality

$$\int_{-\infty}^{\infty} e^{-pt} AY(t)\,dt = \int_{-\infty}^{\infty} e^{-pt} \int_{-\infty}^{\infty} G(t-\tau) Y_1(\tau)\,d\tau = L^{-1}(p) F_1(p)$$

out of the Perceval equality (46) we will receive the following relation

$$\|AY(t)\|^2 = \frac{1}{2\pi} \int_{-\infty}^{\infty} \left| L^{-1}(i\omega) F_1(i\omega) \right|^2 d\omega \le \sup_{\omega} \left| L^{-1}(i\omega) \right|^2 \cdot$$

$$\cdot \frac{1}{2\pi} \int_{-\infty}^{\infty} \left| F_1(i\omega) \right|^2 d\omega = \sup_{\omega} \left| L^{-1}(i\omega) \right|^2 b^2 \cdot \|Y(t)\|^2.$$

Out of it we shall receive the estimation of the norm of the operator A in the space L_2

$$\|A\| \le b \sup_{\omega} \|L^{-1}(i\omega)\|. \tag{48}$$

Finally, we come to the conclusion that for the system of differential-difference equations

$$L(d)Y(t)+\varepsilon B(t)Y(t)=0$$

exponential dichotomy is preserved if the following condition is performed

$$|\varepsilon|\sup_{\omega}|L^{-1}(i\omega)|\cdot\sup_{t}|B(t)|<1. \qquad (49)$$

The received criterion of dichotomy can be applied to find the conditions of keeping stability or instability of the solutions of the equation system

$$L(d)\ Y(t)+\varepsilon B(t)Y(t)=0. \qquad (50)$$

Example. We shall consider the linear differential equation with exponential dichotomy of the solutions

$$y''(t)+vy'(t)-y(t)=0, \qquad (v>0).$$

The solutions of this equation are unstable. We shall find values ε, at which the equation solutions

$$y''(t)+vy'(t)-y(t)+\varepsilon y(t)=0 \qquad (51)$$

will be instable. We shall make the condition (49)

$$|\varepsilon|\cdot\sup_{\omega}\frac{1}{|(\omega i)^{2}+vi\omega-1|}\cdot 1<1$$

and find inequality $|\varepsilon|<1$. At this condition the solutions of the equation (51) are unstable, as the coefficient at y $(\varepsilon-1)<0$.

In the same way it is possible to prove that the zero solution of the equation
$$y''(t)+vy'(t)-y(t)+\varepsilon \sin \omega t\, y(t)=0$$
is unstable at $|\varepsilon|<1$.

We shall show some criteria of stability and instability of the system of differential-difference equations.

I. If the solutions of the system of differential-difference equations
$$L(d)Y(t)=0, \quad \left(d\equiv\frac{d}{dt}\right) \tag{52}$$
are stable (unstable), then when the following condition is performed
$$\sup_{\omega}|L^{-1}(i\omega)|\cdot\sup_{t}|B(t)|<1 \tag{53}$$
the solutions of the equation system
$$L(d)Y(t)+B(t)Y(t)=0$$
are also stable (unstable).

II. If the solutions of the equations system (52) are stable (unstable), then the solutions of the equations system

$$L(d)Y(t) + \sum_{S=1}^{N} L_S(d) B_S(t) Y(t) = 0$$

remain stable (unstable), if the frequency condition of keeping the solution dichotomy is performed.

$$\sum_{S=1}^{N} \sup_{\omega} \left| L^{-1}(i\omega) L_S(i\omega) \right| \cdot \sup_{t} \left| B_S(t) \right| < 1. \qquad (54)$$

III. We shall consider the following system of the differential-difference equations

$$L(d)Y(t) + \sum_{S=1}^{N} B_S(t) L_S(d) Y(t) = 0. \qquad (55)$$

We shall introduce the new vector $Z(t)$

$$Z(t) = L(d)Y(t), \quad Y(t) = L^{-1}(d)Z(t).$$

The system of equations (55) can be written in the form of the following system

$$Z(t) + \sum_{S=1}^{N} B_S(t) L_S(d) L^{-1}(d) Z(t) = 0.$$

Let the solutions of the equations system (52) be stale (unstable). Then the solutions of the equations system (55) will be stable (unstable), if the following inequality is performed

$$\sum_{S=1}^{N} \sup_{\omega} \left| L_S(i\omega) L^{-1}(i\omega) \right| \cdot \sup_{t} \left| B_S(t) \right| < 1. \quad (56)$$

IV. If the solutions of the system of differential-difference equations

$$L_1(d) L_2(d) Y(t) = 0$$

are stable (unstable), then the solutions of the following system of equations

$$L_1(d) L_2(d) + L_3(d) B(t) L_4(d) Y(t) = 0 \quad (57)$$

will be stable (unstable), if the following condition is performed

$$\sup_{\omega} \left| L_1^{-1}(i\omega) L_3(i\omega) \right| \cdot \sup_{t} \left| B(t) \right| \cdot \sup_{\omega} \left| L_4(i\omega) L_2^{-1}(i\omega) \right| < 1$$

.

Example. We shall find the condition of asymptotic stability of the solutions of the differential equation

$$\frac{d^2 y(t)}{dt^2} + 2v \frac{dy(t)}{dt} + v^2 y(t) + \frac{d}{dt} b(t) \frac{dy(t)}{dt} = 0.$$

We shall write the equation in the operational form

$$(d+v)\ (d+v)\ y(t) + d\, b(t)\ d\, y(t) = 0.$$

From the condition of stability in the form of (58)

$$\sup_{\omega}\left|\frac{i\omega}{i\omega+v}\right|\cdot\sup_{t}|b(t)|\cdot\sup_{\omega}\left|\frac{i\omega}{i\omega+v}\right|<1 \qquad (58)$$

we shall find the sufficient condition of stability $\sup_{t}|b(t)|<1$.

For the differential equations

$$\frac{d^2y(t)}{dt^2}+2v\frac{dy(t)}{dt}+v^2y(t)+\frac{d}{dt}b(t)y(t)=0,$$

$$\frac{d^2y(t)}{dt^2}+2v\frac{dy(t)}{dt}+v^2y(t)+b(t)\frac{dy(t)}{dt}=0$$

the asymptotic condition of stability will assume the view of the following inequality

$$\sup_{t}|b(t)|<v.$$

For the differential equation

$$\frac{d^2y(t)}{dt^2}+2v\frac{dy(t)}{dt}+v^2y(t)+b(t)y(t)=0$$

the condition of asymptotic stability will be reduced to the following inequality

$$\sup_{t}|b(t)|<v^2.$$

§6 Criteria for keeping Exponential Dichotomy

We shall consider the differential-difference system of equations

$$L(d)Y(t)=0, \quad \left(d \equiv \frac{d}{dt}\right). \tag{59}$$

Particular solutions of the system of the equations (59) are found out from the equation system

$$L(d)\ e^{pt}C = 0 \quad C = const \tag{60}$$

or

$$L(p+d)C = 0, \quad d \equiv \frac{d}{dt}.$$

Finally we come to the equation for characteristic indicators p

$$\det L(p) = 0. \tag{61}$$

Any root of the equation (61) is complied with the particular solution (60) of the equation system (59).

Let the straight line $\operatorname{Re} p = \sigma$ have no roots of the equation (61), but there are some roots of the equation (61), which are located further to the right from this straight line. Then there is exponential dichotomy of the equation system (59) relatively the indicator σ. The numerous system solutions can be divided into two kinds of varieties. The solutions of the first variety grow twice as fast as $\exp\{\sigma t\}$. The solutions of the second variety grow more slowly than $\exp\{\sigma t\}$ at $t \to +\infty$ and faster than $\exp\{\sigma t\}$ at $t \to -\infty$.

Let $Y(t)$ be a n-dimensional vector with $t \in (-\infty, \infty)$. We shall introduce the norm for the vector $Y(t)$ by the formula

$$\|Y(t)\|_\sigma^2 = \int_{-\infty}^{\infty} \sum_{k=1}^{n} |y_k(t)|^2 e^{-2\sigma t} dt, \qquad (62)$$

where $y_k(t)$ is a k- projection of the vector $Y(t)$.

If $\|Y(t)\|_\sigma < \infty$, then we write $Y(t) \in L_{\sigma 2}$. We denote the Euclidean norm of the vector as $|Y(t)|$

$$|Y(t)|^2 = \sum_{k=1}^{n} |y_k(t)|^2.$$

We have the following equation

$$\|Y(t)\|_\sigma^2 = \int_{-\infty}^{\infty} |Y(t)|^2 \, dt.$$

We shall define the norm of the matrix $B(t)$ in the Euclidean space as $|B(t)|$.

We shall consider the system of differential-difference equations

$$L(d)Y(t) + \varepsilon B(t)Y(t) = X(t),$$
$$|B(t)| \le b, \quad X(t) \in L_{\sigma 2} \qquad (63)$$

We shall suppose that at $\varepsilon = 0$ for the solution of the equation system (63) there is exponential dichotomy of the solutions with the indicator σ, i.e.

$$\det L(\sigma + i\omega) \ne 0, \quad \omega \in (-\infty, \infty).$$

The equation (63) for a particular solution $Y(t) \in L_{\sigma 2}$, will assume the following view

$$Y(t) = X_1(t) - \varepsilon \int_{-\infty}^{\infty} G_\sigma(t-\tau) B(\tau) Y(\tau) \, d\tau,$$

$$X_1(t) \in L_{\sigma 2}, \qquad (64)$$

where it is defined as follows

$$G_\sigma(t-\tau) = \frac{1}{2\pi i} \int_{\sigma - i\infty}^{\sigma + i\infty} L^{-1}(p) \, e^{p(t-\tau)} dp. \qquad (65)$$

We shall use the Perceval formula. If

$$F(p) = \int_{-\infty}^{\infty} Y(t)\, e^{-pt} dt, \quad Y(t) \in L_{\sigma 2},$$

then the Perceval equality is true

$$\int_{-\infty}^{\infty} e^{-2\sigma t} |Y(t)|^2 \, dt = \frac{1}{2\pi} \int_{-\infty}^{\infty} |F(\sigma + i\omega)|^2 \, d\omega. \quad (66)$$

We shall take the integral operator A, which is defined by the expression below, into our consideration

$$AY(t) = \int_{-\infty}^{\infty} G_\sigma(t-\tau) B(\tau) Y(\tau) \, d\tau.$$

Let $Y(t) \in L_{\sigma 2}$, i.e. the following equality is true

$$\|Y(t)\|_\sigma^2 = \int_{-\infty}^{\infty} e^{-2\sigma t} \|Y(t)\|^2 \, dt \le c^2.$$

Then as the vector-function

$$Y_1(t) = B(t) Y(t)$$

has an estimation for the norm in $L_{\sigma 2}$

$$\|B(t)Y(t)\|_\sigma^2 = \int_{-\infty}^{\infty} e^{-2\sigma t} |B(t)|^2 \cdot |Y(t)|^2 \, dt \le b^2 \|Y(t)\|_\sigma^2.$$

Thus, we have got a simple relation

$$\|B(t)Y(t)\|_\sigma \le b \|Y(t)\|_\sigma.$$

We shall introduce the Laplace transformation for $Y_1(t)$

$$F_1(p) = \int_{-\infty}^{\infty} e^{-pt} Y_1(t) dt.$$

According to the convolution theorem we have the equality

$$\int_{-\infty}^{\infty} e^{-pt} AY(t) dt = L^{-1}(p) F_1(p).$$

The Perceval formula (66) gives the relation

$$\|AY(t)\|_\sigma^2 = \frac{1}{2\pi} \int_{-\infty}^{\infty} \left| L^{-1}(\sigma+i\omega) F_1(\sigma+i\omega) \right|^2 d\omega \leq$$

$$\leq \sup_\omega \left| L^{-1}(\sigma+i\omega) \right|^2 \cdot \frac{1}{2\pi} \int_{-\infty}^{\infty} \left| F_1(\sigma+i\omega) \right|^2 d\omega = \sup_\omega \left| L^{-1}(\sigma+i\omega) \right|^2 b^2 \|Y(t)\|_\sigma^2$$

Here we can find the estimation of the norm of the operator A in the space $L_{\sigma 2}$

$$\|A\|_\sigma \leq b \sup_\omega \left| L^{-1}(\sigma+i\omega) \right|. \qquad (67)$$

The system of the integral equations (64) in the space $L_{\sigma 2}$ can be written in the form

$$Y(t) = X_1(t) - \varepsilon AY(t)$$

and according to the Bannah theorem it can be solved by the method of sequential approximations under the condition

$$|\varepsilon| \cdot \|A\|_\sigma < 1$$

or

$$|\varepsilon|\ b\sup_{\omega}|L^{-1}(\sigma+i\omega)|<1. \tag{68}$$

The solution to be found is $Y(t)\in L_{\sigma 2}$.

We come to the conclusion which states that for the system of differential-difference equations (63) exponential dichotomy of the solutions is kept relatively the indicator σ, if the following condition is performed

$$|\varepsilon|\ \sup_{\omega}|L^{-1}(\sigma+i\omega)|\cdot\sup_{t}|B(t)|<1.$$

The received criterion of dichotomy can be applied for finding conditions of existence of the solutions which are growing at $t\to+\infty$ more slowly or faster than $\exp\{\sigma t\}$.

Example. Solutions of the differential equation

$$y''(t)-2y'(t)=0$$

have dichotomy with the indicator $\sigma=1$. Under this the roots of the characteristic equation $p_1=0$, $p_2=2$ are located on different sides of the straight line $\mathrm{Re}\,p=1$.

For the differential equation

$$y''(t)-2y'(t)+\varepsilon\sin t\cdot y(t)=0$$

the condition of preservation of dichotomy (68) will assume the following view

$$|\varepsilon| \cdot 1 \cdot \sup_{\omega} \left| \frac{1}{(1+i\omega)^2 - 2(1+\omega)} \right| < 1$$

and it is converged to the inequality $|\varepsilon| < 1$.

We shall demonstrate frequency criteria of preservation of the exponential dichotomy when the system of differential-difference equations is disturbed.

I. Let for the system of differential-difference equations

$$L(d)Y(t) = 0, \quad d \equiv \frac{d}{dt} \tag{69}$$

there be exponential dichotomy of solutions with the indicator σ.

When the following condition is

$$|\varepsilon| \sup_{\omega} \left| L^{-1}(\sigma + i\omega) \right| \cdot \sup_{t} |B(t)| < 1 \tag{70}$$

exponential dichotomy of the solution of the equation system

$$L(d)Y(t) + \varepsilon B(t)Y(t) = 0$$

is preserved.

II. In the same way for the system of equations

$$L(d)y(t)+\varepsilon\sum_{S=1}^{N}L_S(d)B_S(t)Y(t)=0$$

exponential dichotomy with the indicator σ is preserved if the following condition is fulfilled

$$|\varepsilon|\cdot\sum_{S=1}^{N}\sup_{\omega}|L^{-1}(\sigma+i\omega)L_S(\sigma+i\omega)|\cdot\sup_{t}|B_S(t)|<1. \quad (71)$$

III. Let for the system of equations (69) there be exponential dichotomy of the solutions with the indicator σ.

For the system of equations

$$L(d)Y(t)+\varepsilon\sum_{S=1}^{N}B_S(t)L_S(d)Y(t)=0$$

exponential dichotomy of solutions with the indicator σ is preserved if the following inequality if performed

$$|\varepsilon|\sum_{S=1}^{N}\sup_{\omega}|L_S(\sigma+i\omega)L^{-1}(\sigma+i\omega)|\cdot\sup_{t}|B_S(t)|<1. \quad (72)$$

IV. Let for the solutions of the system of differential-difference equations

$$L_1(d)L_2(d)Y(t)=0$$

there be exponential dichotomy of solutions with the indicator σ. If the following inequality is performed

$$|\varepsilon|\sup_{\omega}|L_1^{-1}(\sigma+i\omega)L_3(\sigma+i\omega)|\cdot$$
$$\cdot\sup_t|B(t)|\cdot\sup_{\omega}|L_4(\sigma+i\omega)L_2^{-1}(\sigma+i\omega)|<1 \quad (73)$$

then exponential dichotomy with the indicator σ is preserved for the solutions of the equation system

$$L_1(d)L_2(d)Y(t)+\varepsilon L_3(d)B(t)L_4(d)Y(t)=0.$$

§7   Method of the Transfer Function

The notion of a transfer function plays a big role in the automatic control theory. Nowadays is mainly applied for the research of stationary dynamic systems.

At first we shall consider stationary systems of differential equations. We can often come across the linear differential equation with constant coefficients in applied problems

$$\sum_{k=0}^{n} a_k \frac{d^k y(t)}{dt^k} = \sum_{k=0}^{m} b_k \frac{d^k x(t)}{dt^k} \quad (m \leq n). \quad (74)$$

The value $x(t)$ is called input or an influence at the input. The value $y(t)$ is called output or a reaction at the output. In the simplest case when the input is defined by a demonstrative function

$$x(t) = e^{pt} \quad (p = const),$$

the output can be found in the same form

$$y(t) = H(p)\, e^{pt}. \quad (75)$$

If such a particular solution exists, then the depending on p relation of the output to the input

$$H(p) = \frac{y(t)}{x(t)}$$

is called a transfer function of the equation (74). We shall find the analytical expression for the function $H(p)$

$$H(p) = \frac{B(p)}{A(p)}, \quad A(p) = \sum_{k=0}^{n} a_k p^k,$$

$$B(p) = \sum_{k=0}^{m} b_k p^k, \qquad (76)$$

which is fractionally rational function. Poles $A(p)$ are located in the points $p = p_k$ $(k = 1,...,n)$, where p_k are characteristic indicators of the equation solutions (74) at $x(t) \equiv 0$, which are defined by the following equation

$$A(p) = 0.$$

We shall introduce the image of the function $x(t)$, $y(t)$ by Laplace

$$f_1(p) = \int_0^{\infty} e^{-pt} x(t) \, dt, \quad f_2(p) = \int_0^{\infty} e^{-pt} y(t) \, dt.$$

Let the zero initial conditions for the functions $x(t)$, $y(t)$ are fulfilled at the point $t = 0$

$$x(0)=0,\ \frac{dx(0)}{dt}=0,...,\ \frac{d^{m-1}x(0)}{dt^{m-1}}=0,$$

$$y(0)=0,\ \frac{dy(0)}{dt}=0,...,\ \frac{d^{n-1}y(0)}{dt^{n-1}}=0$$

The differential equation (74) will change into the equation for the images

$$\sum_{k=0}^{n} a_k p^k f_2(p) = \sum_{k=0}^{m} b_k p^k f_1(p).$$

Here we can see that the transfer function $H(p)$ can be defined as a relation of the image $f_2(p)$ of the system reaction to the image $f_1(p)$ of the effect at the input at the zero values of the output and input

$$H(p) = \frac{f_2(p)}{f_1(p)}. \qquad (77)$$

The fact that the transfer function can be defined experimentally is more essential in the theory of using. Usually the coefficients a_k, b_k in the equation (74) can be partly and the formula (76) is practically improper for finding $H(p)$. If all the roots of the polynomial $A(p)$ have negative real parts, then we can find coefficients of the transfer function approximately. One of the ways to do it looks as follows: a simple harmonical action is applied to the input

$$x(t) = \sin \omega t.$$

At the output of the system after finishing the transfer process we will receive an established harmonical reaction

$$y(t) = \beta(\omega)\sin(\omega t + \alpha(\omega)) \qquad (\beta(\omega) > 0).$$

The functions $\alpha(\omega), \beta(\omega)$ are found experimentally and called frequency and amplitude characteristics of the system (74).

If at the input we cause a complex influence

$$x(t) = e^{i\omega t} = \sin\left(\omega t + \frac{\pi}{2}\right) + i\sin \omega t,$$

then, as the system (74) is linear, we shall receive the reaction at the output

$$y(t) = \beta(\omega)\sin\left(\omega t + \alpha\left(\omega + \frac{\pi}{2}\right)\right) + i\beta(\omega)\sin(\omega t + \alpha(\omega)) =$$

$$= \beta(\omega)\, e^{i(\omega t + \alpha(\omega))}.$$

Out of it we shall receive am important relation for the transfer function

$$H(i\omega) = \beta(\omega)e^{i\alpha(\omega)}.$$

Knowing the $H(i\omega)$, we can define the $H(p)$. The approximate methods of building the transfer function $H(p)$ correspond to the approximate methods of defining $\alpha(\omega)$, $\beta(\omega)$. The function $H(p)$ is analytical, so the function $\ln H(p) = \ln|H(p)| + i\arg H(p)$ is also analytical. At $p = i\omega$ the function is

$$\ln H(i\omega) = \ln|\beta(\omega)| + i\alpha(\omega).$$

Knowing only the real or the imaginary part $\ln H(i\omega)$ we can approximately restore the functions $\ln H(p)$, $H(p)$ [17].

Another important method of finding the transfer function is connected with using of the Laplace conversion. Let $f_1(p) \equiv 1$, i.e. the input influence $x(t)$ is a generalized Dirac δ-function, a mathematical model of the impact influence. The function $\delta(t)$ is defined by the formulas

$$\delta(t) = 0 \quad (t \neq 0), \quad \delta(0) = +\infty,$$

$$\int_{-a}^{a} f(t)\delta(t)dt = f(0), \quad (a > 0).$$

In this case the image $f_2(p)$ of the solution $g(t)$ of the linear differential equation

$$\sum_{k=0}^{n} a_k d^k g(t) = \sum_{k=0}^{m} b_k d^k \delta(t), \quad \left(d \equiv \frac{d}{dt}\right) \quad (78)$$

will coincide with the transfer function $H(p)$. The solution $y = g(t)$ itself of the equation (78) is called an impulse transfer function. It can be received experimentally making an impact influence at the input of the system (74), which stays quiescent at $t < 0$. If the function $y = g(t)$ is found experimentally or by a calculating way then the particular solution of the equation (74) can be found at zero initial values for $y(t)$, $x(t)$ with the help of the convolution formula

$$y(t) = \int_0^t g(t-\tau) \cdot x(\tau) d\tau.$$

It is apparent that other kinds input influences can be applied, for example a graded one

$$x(t) = 0 \quad (t < 0), \quad x(t) = 1 \quad (t > 0),$$
$$f_1(p) = p^{-1}.$$

The knowledge of a transfer function is equal to the knowledge of coefficients of the equation system (74). For the system of linear differential equations

$$\frac{dY(t)}{dt} + AY(t) = X(t), \quad A = const,$$

the analogue of the transfer function will be the transfer matrix
$$H(p) = (Ep + A)^{-1}.$$

If $A(d)$, $B(d)$ are matrix differential operators, then in the common case the transfer operators from $X(t)$ to $Y(t)$ for the equation system
$$A(d)Y(t) = B(d)X(t), \quad \left(d = \frac{d}{dt}\right),$$
will be the following matrix
$$H(p) = A^{-1}(p)B(p).$$

Example. We shall consider the differential-difference equation
$$\frac{dy(t)}{dt} + y(t-\tau) = x(t) - x(t-\tau),$$
which can be written in the operational form
$$\left(d + e^{-d\tau}\right) y(t) = \left(1 - e^{-d\tau}\right) x(t).$$

The transfer function assumes the following view
$$H(p) = \left(1 - e^{-p\tau}\right) \left(p + e^{-p\tau}\right)^{-1}.$$

§8 The Transfer Function of the Non-stationary System of Differential Equations

We shall consider a linear differential equation with variable coefficients

$$\sum_{k=0}^{n} a_k(t)\frac{d^k y(t)}{dt^k} = \sum_{k=0}^{m} b_k(t)\frac{d^k x(t)}{dt^k}, \qquad (79)$$

which can be written in the operational form

$$A(t,d)y(t) = B(t,d)\,x(t),$$

where it is defined as

$$A(t,d) = \sum_{k=0}^{n} a_k(t)d^k, \quad B(t,d) = \sum_{k=0}^{n} b_k(t)d^k,$$

$$\left(d \equiv \frac{d}{dt}\right).$$

We shall suppose that the influence on the input $x(t)$ has a special view

$$x(t) = e^{pt} \qquad (80)$$

and we shall find the corresponding particular solution of the equation (79) which looks as the follows

$$y(t) = e^{pt} w(t,p).$$

The function $w(t,p)$ can be used for the solution of the equation (79) at the random influence o the input. Namely, let the function $x(t)$ be introduced with the help of the Laplace inversion according to the formula

$$x(t) = \frac{1}{2\pi\varepsilon} \int_{\sigma-i\infty}^{\sigma+i\infty} e^{pt} \psi(p)\, dp,$$

where σ is a big enough number, and $\psi(p)$ is a display og the function $x(t)$

$$\psi(p) = \int_0^\infty e^{-pt} x(t)\, dt.$$

We shall multiply the equation

$$A(t,d) e^{pt} w(t,p) = B(t,d)\ e^{pt} \quad \left(d = \frac{d}{dt}\right) \quad (81)$$

by the function на $\psi(p)$ and integrate by p along the straight line $\operatorname{Re} p = \sigma$. Supposing the permutability of differentiation operations by t and integration by p, we come to the equation

$$A(t,d)\frac{1}{2\pi i}\int_{\sigma-i\infty}^{\sigma+i\infty} e^{pt}w(t,p)\ \psi(p)\ dp = B(t,d)\ x(t).$$

This equation follows that in the equation (79) the written bellow expression can be taken as $y(t)$

$$y(t) = \frac{1}{2\pi i}\int_{\sigma-i\infty}^{\sigma+i\infty} e^{pt}w(t,p)\ \psi(p)dp. \qquad (82)$$

This formula shows that it is sufficient to solve the equation (79) at the special influence (80) at the in order to be able to solve the equation (79) at the random influence $x(t)$.

The equation (81) for finding the transfer function $w(t,p)$ can be transformed into another form. Using the well-known in the differential equation theory of operators formula

$$A(t,d)\ e^{pt}w(t,p) = e^{pt}A(t,p+d)\ w(t,p)$$

we shall come to the equation

$$e^{pt}A(t,p+d)\ w(t,p) = e^{pt}B(t,p+d),$$

or

$$A(t,p+d)\ w(t,p) = B(t,p), \qquad d = \frac{\partial}{\partial t}. \quad (83)$$

In our case the expression $A(t,d)$ is a polynomial relatively d. This polynomial can be expressed as a power series in d by Taylor series expansion

$$A(t,p+d) = \sum_{k=0}^{n} \frac{1}{k!} \frac{\partial^k A(t,p)}{\partial p^k} \cdot d^k$$

and rewrite the equation (83) as, [5]

$$\sum_{k=0}^{n} \frac{1}{k!} \frac{\partial^k A(t,p)}{\partial p^k} \cdot \frac{\partial^k w(t,p)}{\partial t^k} = B(t,p). \qquad (84)$$

We shall show that this equation can be further simplified. To do this we shall come to the formula (82) in the space of originals. We shall introduce the function of two variables

$$g(t,t-s) = \frac{1}{2\pi i} \int_{\sigma-i\infty}^{\sigma+i\infty} w(t,p) \; e^{ps} dp. \qquad (85)$$

The formula (82), if we replace s by $t-\tau$, will assume to following look

$$y(t) = \int_0^t g(t,\tau) \; x(\tau) \; d\tau \qquad (86)$$

and it is known as the well-known Cauchy formula for building a particular solution of the inhomogeneous equation (78) with the zero initial conditions at $t=0$.

Is the function $g(t,\tau)$ is known, then the parametric transfer function $w(t,p)$ can be fond with the help of the Laplace transformation

$$w(t,p) = \int_0^\infty g(t,\tau) \ e^{-p(t-\tau)} d\tau = \int_\tau^\infty g(t,\tau) \ e^{-p(t-\tau)} d\tau.$$

Let a "simpler" than the (84) equation be solved

$$\sum_{k=0}^n \frac{1}{k!} \frac{\partial^k A(t,p)}{\partial p^k} \cdot \frac{\partial^k \upsilon(t,p)}{\partial t^k} = 1. \qquad (87)$$

The corresponding Cauchy function is denoted via $h(t,\tau)$

$$h(t,\tau) = \frac{1}{2\pi i} \int_{\sigma-i\infty}^{\sigma+i\infty} \upsilon(t,p) \ e^{p(t-\tau)} dp.$$

Then the solution of the equation (84) can assume the following view

$$y(t) = \int_0^t h(t,\tau) \sum_{k=0}^m b_k(\tau) \frac{d^k x(\tau)}{d\tau^k} d\tau.$$

Integrating by parts and supposing that the outside of integral members go to zero because of choosing zero initial conditions for $x(t)$ in the point $t=0$, we come to the following formula

$$y(t) = \int_0^t \sum_{k=0}^m (-1)^k \frac{\partial^k}{\partial \tau^k} \left[h(t,\tau) \ b_k(\tau) \right] x(\tau) d\tau.$$

Comparing with the formula (86) gives the equality

$$g(t,\tau) = \sum_{k=0}^{m}(-1)^k \cdot \frac{\partial^k}{\partial \tau^k}\left[h(t,\tau)\ b_k(\tau)\right]. \quad (88)$$

Thus, it is sufficient to solve the equation (87), in order to solve the equation (79). Inversely, if we can solve the initial equation (79), we can solve the "more particular" equation (84) or (87). There a well-known cases in mathematics when a particular case is equivalent to the common one. In this case the solution of the equation (79) and the solution of the equation (87) are equivalent tasks and have similar difficulty. The advantages of the introduced notions make formalization of the known formulas more convenient, which is conditioned by the conventional for engineers using the Laplace transformation. For the equation (87) we can develop approximate methods of the solution. The hidden advantages make it possible to use the function features of a complex variable for presenting and analytical continuing of the function $v(t,p)$. The disadvantage is that the function $w(t,p)$ is multi-valuedly defined by the equation (87).

The main conclusion states that for building the parametric transfer function $w(t,p)$ and for the equation (79) we need to be able to integrate this equation in the closed and approximate form.

Example. We shall find a parametric transfer function for the differential equation.

$$\frac{dy(t)}{dt} + ñht \cdot y(t) = x(t).$$

Preliminarily we shall find the Cauchy function

$$g(t,\tau) = \frac{ch\tau}{cht} \quad (t<\tau), \quad g(t,\tau) = 0 \quad (t>\tau),$$

and then the parametric transfer function

$$w(t,p) = \int_t^\infty \frac{ch\tau}{cht} e^{-p(t-\tau)} d\tau = \frac{1}{p+1} + \frac{e^{-t}}{cht} \cdot \frac{1}{p^2-1}.$$

At quite big values t the function $w(t,p)$ is approximate to the function

$$w_1(p) = \frac{1}{p+1}$$

i.e. at quite big values t the initial differential equation can be approximately replaced by the equation with constant coefficients

$$\frac{dy(t)}{dt} + y(t) = x(t).$$

§9 Generalization of the transfer function notion

Let $L(t,d)$ be a differential-difference operator, $\left(d \equiv \dfrac{d}{dt}\right)$. If for any bounded piecewise-continuous vector-function $X(t)$, which satisfies the restriction

$$\left|e^{-\sigma t} X(t)\right| \leq c = const, \qquad -\infty < t < \infty \qquad (89)$$

the system of linear equations

$$L(t,d)\ Y(t) = X(t)$$

always has a single solution $Y(t)$, which satisfies the condition of restriction

$$\left|e^{-\sigma t} Y(t)\right| \leq mc,$$

then this solution can be presented as

$$Y(t) = \int_{-\infty}^{\infty} G_\sigma(t,\tau) X(\tau) d\tau, \qquad (90)$$

where the following inequality is performed

$$\int_{-\infty}^{\infty} e^{-\sigma(t-\tau)} \left|G_\sigma(t,\tau)\right| d\tau \leq m.$$

The operator $G_\sigma(t,\tau)$ will be called the Green parametric operator. The existence of the operator $G_\sigma(t,\tau)$ implies the existence of the parametric transfer operator $W(t,p)$

$$W(t,p) = \int_{-\infty}^{\infty} G_\sigma(t,\tau) \ e^{-p(t-\tau)} d\tau,$$

$$G_\sigma(t,\tau) = \frac{1}{2\pi i} \int_{\sigma-i\infty}^{\sigma+i\infty} W(t,p) \ e^{p(t-\tau)} dp, \qquad (91)$$

which is holomorphic on the straight line $\operatorname{Re} p = \sigma$. Teen operator he existence of the Green operator $G_\sigma(t,\tau)$ implies an exponential dichotomy of the solutions of the homogeneous system of equations

$$L(t,d)Y(t) = 0 \qquad (92)$$

relatively the indicator σ. At this there are the such constants $q > 0$, $\delta > 0$ that

$$\left| G_\sigma(t,\tau) \ e^{-\sigma(t-\tau)} \right| \le q e^{-\delta|t-\tau|}.$$

Any solution $Y(t)$ of the systems of equations(92) is uniquely divided into two components

$$Y(t) = Y_1(t) + Y_2(t),$$

each of which satisfies the following conditions:

$$|Y_1(t)e^{-\sigma t}| \le q_1 e^{-\delta t} \quad (t>0),$$
$$|Y_2(t)e^{-\sigma t}| \le q_2 e^{\delta t} \quad (t<0).$$

We shall note that in the stationary case for the equation

$$L(d)\ Y(t) = X(t)$$

the Green parametric operator is defined by the formula

$$G_\sigma(t,\tau) = \frac{1}{2\pi i}\int_{\sigma-i\infty}^{\sigma+i\infty} L^{-1}(p)\ e^{p(t-\tau)}dp. \qquad (93)$$

Commonly, at rather high values σ the Green operator turns into the Cauchy operator and satisfies the condition

$$G_\sigma(t,\tau) \equiv 0 \quad (\tau > t).$$

§10 Successive approximation method

We shall make an integral equation for the Green operator $G_\sigma(t,\tau,\varepsilon)$ of the system of differential-difference equations

$$L(d)Y(t) + \varepsilon L_1(t,d)Y(t) = 0, \qquad d = \frac{d}{dt}. \quad (94)$$

Using the formula (93), we shall receive the integral equation

$$G_\sigma(t,\tau,\varepsilon) = G_\sigma(t,\tau,0) - \varepsilon \int_{-\infty}^{\infty} G_\sigma(t,s,0)\ L_1(s,d)\ G_\sigma(s,\tau,\varepsilon)\ ds,$$

where the operator $L_1(s,d)$ influences on the following expression on the variable s $\left(d = \dfrac{\partial}{\partial s}\right)$. With the help of an integration by pieces and argument shift formula we shall transform this operational integral equation into the Fredgolm integral equation

$$G_\sigma(t,\tau,\varepsilon) = G_\sigma(t,\tau,0) + \varepsilon \int_{-\infty}^{\infty} R_\sigma(t,s) G_\sigma(s,\tau,\varepsilon) ds, \quad (95)$$

where we shall introduce a definition for the nucleus

$$R_\sigma(t,s) = -G_\sigma(t,s,0)L_1(s,-d) \quad \left(d = \frac{\partial}{\partial s}\right)$$

and the operator $L_1(s,-d)$ influences on the previous expression. The integral equation (95) can be solved with the help of successive approximation method which converges if the following condition is performed

$$|\varepsilon|\int_{-\infty}^{\infty} e^{-\sigma(t-s)}|R_\sigma(t,s)|\ ds \leq l < 1. \tag{96}$$

For the transfer operator $W(t,p,\varepsilon)$

$$W(t,p,\varepsilon) = \int_{-\infty}^{\infty} G_\sigma(t,\tau,\varepsilon)\ e^{-p(t-\tau)}d\tau$$

the equation (96) complies with the operator integral equation

$$W(t,p,\varepsilon) = L^{-1}(p) + \varepsilon\int_{-\infty}^{\infty} R_\sigma(t,\tau)\ e^{-p(t-\tau)}W(\tau,p,\varepsilon)d\tau \tag{97}$$

Is the condition (96) is performed the exponential dichotomy of the equation solution (94) relatively the index σ is preserved. A variety of solutions of the equation system (94) at the exponential dichotomy can be parted into two integral varieties, whose dimension does not change if the following condition is performed (96). If $\sigma = 0$ and all solutions of the system solutions (94) are asymptotically stable at $\varepsilon = 0$ at $t \to +\infty$, then they will remain asymptotically stable if the condition (96) is performed.

We shall point to the perspectives of the transfer function method development. Currently the linear differential equation theory has a formal character. It deals with abstract notions which can not be calculated in the general view at present. For example Lyapunov indicators for the solution can be found approximately only for some simple classes of linear differential equations. Thus the general theory of linear differential equations has a more qualitative than quantitative character. Using of transfer functions gives a hope for creation of analytical apparatus for approximate calculation of Lyapunov indicators.

The present theory of linear differential equations is usually applied for the system of equations of the first order. In the engineering practice we often come across equations of the second order and higher. The theory of such equations is developed mainly application engineers. It resulted in intrusion into the theory of generalized functions. It is especially easy to formalize these calculations with the help of the Laplace transformation, which makes it expedient to use the transfer function method.

The parametric transfer function analytically depends on the parameter, which enables to use the ideas of analytical continuing on this parameter. Presently the transfer function method is far from completion. The calculating methods of building are not developed enough. When this method spreads on difference equations and equations with partial derivatives we come across difficulties connected with using generalized functions. We can suppose that development of the transfer function method will be connected with specificity of the coefficients. In particular, the most profound results were received for the system of linear differential equations with periodic coefficients [6]. On the whole, the transfer function method enables to research various results of the general theory of linear differential equations.

Tasks for solving.

1. Receive frequency criteria for stability for difference equations using z-transformation.
2. Prove the validity of frequency criteria for stability (58) by other methods, for example by the Lyapunov function method.
3. Find the conditions, at performing which, the indicators of the exponential growth of the solutions of the differential-difference equation are located in the given interval $\sigma_1 < \operatorname{Re} p < \sigma_2$.
4. Apply the transfer function method for solution and research of the system of differential-difference equations like
$$L(d)\ Y(t) + \varepsilon L_1(t,d)\ Y(t) = 0,$$
where ε is a small parameter.
5. Study the systems of equations like
$$L(\tau,d)Y(t) = 0, \qquad \tau = \varepsilon \cdot t.$$
Study the cases of adhesion and intersection of roots $p_j(\tau)$ of the eqaution $\det L(\tau, p) = 0$.

6. Connect the theory of characteristic numbers of Lyapunov with asymptotic display of a parametric transfer function.

Literature

1. Babakov N.A., Voronov N.A., Voronova A.A. and others. Theory of automatic control. Part 1. Theory of linear systems of automatic control. – M.: High school, 1986. – 367 p.

2. Valeev K.G. Transfer function method. – Kiev, Society "Znanie", Ukrainian SSR, 1977. – 30 p.

3. Eisermann M.A., Handmacher F.R. Absolute stability of the regulated systems. – M.: Publishing Academy of Science USSR, 1963.

4. Nelepyn R.A. (edit.) Methods of non-linear systems of automatic control research. – M.: "Nauka", 1975.

5. Zadeh L.A. Proc. IRE, 38, №3, 1950.

6. Shilman S.V. Method of generating functions in the differential system theory. – M.: Nauka, 1978. – 235 c.

7. Solodov V.A. Linear systems of automatic control with variable parameters. – M., Phismatgyz, 1962.

8. Mihaylov F.A., Teryaev E.D., Bulgakov V.P., Salikov L.M., Dikanova L.S. Dynamics of continuous liner systems with determinate and random parameter. – M.: "Nauaka", 1971.

9. Rosenwasser E.N. Periodacally non-stationary systems of control. – M.: "Nauka", 1973.

10. Sadovnikov V.V. Statistic dynamics of linear systems of automatic control. – M.: Phismatgyz, 1960.

Project *"Modern Mathematics for Engineers"* includes publishing the following works:

1. Difference equations with random coefficients
2. Stability of solutions of differential equations systems with random coefficients
3. Random values modeling
4. Optimal control synthesis
5. Principle of reduction
6. New method of averaging
7. New determinant theory
8. Minimax criterion of stability
9. Numerical methods of stability research
10. Analytical functions from matrix
11. Frequency criterion of stability

"Modern Mathematics for Engineers"

The Author

Prof. Tamara Stryzhak

Translator

Nataliya Sarycheva

Editors

Mirjana Vukelja (Switzerland, ETH Zurich), Bettina Carina Siebert (Germany, TU Munich), Laura Lambertz (Germany, RWTH Aachen)

First IAESTE trainee students

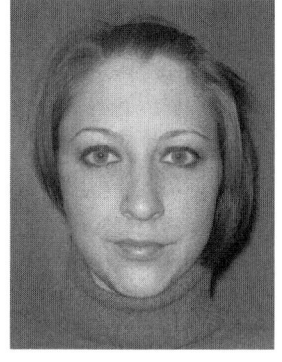

Mirjana Vukelja

Switzerland

ETH Zurich

Bettina Carina Siebert

Germany

TU Munich

Laura Lambertz

Germany

RWTH Aachen

Ki Yeun Kim

USA

Carnegie Mellon University

Department of Mathematical Sciences

Peter Smith

UK

Queen's University Belfast

School of Mathematics and Physics

Heide Gieber

Austria

Vienna University of Technology

Department of Economical Mathematics

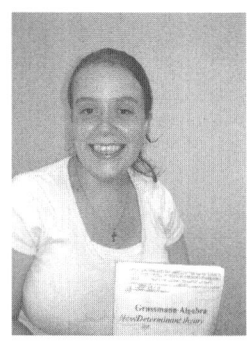

Sarah Burden

UK

University of St. Andrews

Department of Mathematics

(Prof. E. Priest)

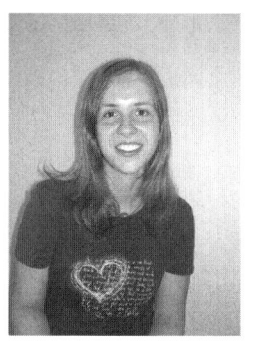

Andrea Trautsamwieser

Austria

Technical University of Vienna

Department of Mathematics

Leonard Neuhaus

Germany

Ludwig - Maximilians University of Munich

Department of Physics

Erkan Koc

Germany

University of Bonn

Department of Computer Science

David Ellis,
Imperial College,
London, UK

Henning Hans Petzka, RWTH Aachen, Germany

Leonard Neuhaus, Sarah Burden, Andrea Trautsamwieser, Erkan Koc

Summary of results from the first volume

Minimax criteria of stability

$$L = T - f$$

$$\max_{q}\left\langle \min_{\dot{q}} L(t,\dot{q}_j,q_j)\right\rangle$$

"Mathematical Truth stays put in centuries, Metaphysical Ghosts disappear like delirium of the sick."

Voltaire

Minimax Criterion of Stability is the first volume within the project *Modern Mathematics for Engineers*.

The project *Modern Mathematics for Engineers* was launched by AUS–DAAD and the National Committee IAESTE Ukraine. The purpose of the project is to publish the most essential mathematical results, to present them with enough clarity for engineers to apply and, ante omnia, to realize the scientific exchange in the sphere of applied mathematics, the oscillation theory, theoretical mechanics, etc.

We expect to realize this purpose first of all with the help of IAESTE trainee students.

We invite for cooperation all mathematicians and engineers who are interested in having their scientific works published in the frame of the project.

We were inspired to launch the project by the positive experience of the researchers of the Californian University, who published the monograph *Modern Mathematics for Engineers*[1] more than half a century ago, in 1956. This work was very successful. As a matter of fact this monograph laid the solid foundation for the applied mathematics to develop successfully and steadily. This monograph is an excellent example to follow.

---

[1] Edited by Edwin F. Beckenbach (Professor of Mathematics, University of California, Los Angeles). New York, Toronto, London: McGraw-Hill 1956.

Thus, like 50 years ago, we are launching the project with researches dedicated to the pendulum. We can hardly remember any other mechanism which is simpler than the pendulum, the mechanism whose scientific life has been so rich in application in different spheres of organic and inorganic nature, as the pendulum has lived a long and rich in discoveries life. I think it is the right time to build a monument to a simple and meaningful device.

Our contribution into realization of this project included taking the following steps:
1) publishing the monograph *Research Methods of the Pendulum Dynamic Systems* [4];
2) receiving "Minimax Criterion of Stability" and applying it in order to research a number of mathematical models and proved, in particular, that any position of the pendulum, even a horizontal one in the vertical plane, can be made stable with the help of the suspension point oscillations;
3) building an installation to demonstrate our theoretical results.

All in all we tried our best to state the results avoiding proving theorems as their translation from Russian into English might have some inadequacy as grammars of these languages have a certain level of ambiguity. Instead of proving the theorems we shall follow Lopital's words, the author of the first textbook on Mathematical Analysis, and say, "We pass the word of honor that the theorem is true".

Oscillations – what are they?

> "Nature prefers oscillating motions in all demonstrations of life. Not without reason we can assume that there is some optimality property at the back of this phenomenon."[8]

It would be rather difficult to mention all publications about the Pendulum. As a matter of fact, oscillations of pendulum systems were researched by Galileo, Newton, Euler, Huygens and many other scientists who made a great contribution into studying the mechanism of the pendulum motion, which was of great importance for the history of mankind's discoveries and technical progress. The pendulum has been used in clocks to define time, in special devices to measure the terrestrial gravitation, as a plumb line in building to define the vertical, etc. The Earth's rotation was proved with the help of the pendulum oscillations as well.

Pendulum systems are, as a rule, non-linear and require specific methods of research. For instance, creation of the elliptical function theory by Abel, Jacoby, and Weierstrass is also connected with the research of the mathematical pendulum oscillations.

Nowadays due to the mathematicization of research in different sciences there has been an increasing interest in studying motions of pendulum systems. The following special terms have appeared: the pendulum law of the population migration, the pendulum law of the rhythm regulator action, the pendulum of emotions, etc.

It is a well-known fact that all living organisms have so called biological clocks, at the basis of which there is an oscillating system – a non-linear oscillator. It is well known that the vestibular apparatus of animals and humans contains three non-linear pendulum systems, which are located in three mutually perpendicular planes. Oscillations are present everywhere: in the opening of a flower at the sunrise, in the growth of an embryo, in germinating of a grain, in the heart beating, in the work of a dental drill and a jackhammer, in the rise and fall of the tide. Besides, wherever there is life, there are oscillations at the cellular level.

The nature of forces which cause oscillations is variable, but the result of these forces is the same, namely - oscillations.

This information bears a descriptive character as we would like to attract the readers' attention to the oscillation theory to convince them of the fact that oscillation processes imply something much deeper than what we actually know: that the source of the oscillating processes is a hidden potential force and kinetic energy.

Oscillations of a non-linear oscillator can be forced or can have a free-running character (such as the heart or aorta beating, biorhythms, etc.). In fact, the pendulum motion laws, periodic or almost-periodic oscillations, are immanent in the whole physical world.

A human being receives the main information about the outer world via sound and light oscillations, analyzing which scientists use pendulum systems. The latter are found in different engineering tasks. For example, oscillations in electrical and mechanical systems, rocking of ships on water, oscillations of satellites, vibration of the hull and the wings of planes, movements of cables, chains, travelling or gantry cranes, etc. All these phenomena are explained with analogous differential equations which describe motions.

At this point our introduction is completed and we turn to the beautiful algorithmic mathematical language: *"A" is given, "B" is to be proved.*

Minimax criteria of stability

The stability of oscillations of a mechanical or electrical system, which is affected by low amplitude high-frequency disturbances, is researched. It is shown that the affect of vibration can be replaced by the action of some potential force. Under the action of some disturbance the stable equilibrium position can become unstable, and the unstable equilibrium position can become stable. New positions of stable dynamic equilibrium can also appear. Conclusions about the stability can be made with the help of the minimax criteria for stability described in this work. The criteria for stability rests on the time averaging operation of the canonical system of differential equations. The minimax criteria for stability is applied when the frequency of external disturbance is substantially higher than the natural vibration of the quiescent system.

In this work the minimax criteria for stability is applied in the research of the stability of the pendulum systems equilibrium position. In particular, it is found that any position of the pendulum, even a horizontal one, can be made stable via vibrations of the pendulum's suspension point.

An experimental installation to check the received theoretical conclusions was created. All the conclusions were proved correct.

This work illustrates simple examples of the minimax criteria for stability application. It also describes the experiment. At the end of this work the mathematical survey of the minimax criteria for stability is presented.

§1 Minimax stability criteria definition

Here the movement of a mechanical or electrical system is studied. We shall define the generalized coordinates by q_j ($j=1,...,n$), the generalized velocities by $\dot{q}_j$ ($j=1,...,n$) and the time by t.

Let $L = L(t, \dot{q}_j, q_j)$ be the kinetic potential, where $L = T - \Pi$, and T is the kinetic energy of the system and Π is the potential energy. We shall find the minimum of the kinetic potential L as a function of the generalized velocities $\dot{q}_j$,

$$\Pi_0(t, q_j) = \min_{\dot{q}_s} L(t, \dot{q}_j, q_j), \quad (j, s = 1,...,n).$$
(1)

We shall exclude from $L(t, \dot{q}_j, q_j)$ the derivative $\dot{q}_j$ by the necessary conditions of extremum

$$\frac{\partial L(t,\dot{q}_j,q_j)}{\partial \dot{q}_s}=0, \quad (j,s=1,...,n).$$
(2)

Using the operation of time averaging t on the function $\Pi_0(t,q_j)$ we are able to eliminate t

$$\Pi_0(q_j) \equiv \langle \Pi_0(t,q_j) \rangle \equiv \lim_{T \to +\infty} \frac{1}{T} \int_0^T \Pi_0(t,q_j)dt.$$
(3)

If the function $\Pi_0(q_j)$ has a maximum at the point $q_j = q_{j0}$ $(j=1,...,n)$, then the maximum corresponds to the stable position of the dynamic equilibrium of the initial oscillating system. Meanwhile the system is steadily oscillating at the position $q_j = q_{j0}$ $(j=1,...,n)$.

To find the position of the stable state it is first necessary to find the minimum of the function $L(t,\dot{q}_j,q_j)$ via variable $\dot{q}_j$, and then find the maximum of the function using,

$$\max_{q_j} \left\langle \min_{\dot{q}_j} L\left(t, \dot{q}_j, q_j\right) \right\rangle,$$

by which the name *'minimax criteria of stability'* arises.

It is important to mention that the conditions of stability are found without using Lagrange's differential equations [1]

$$\frac{d}{dt}\frac{\partial L}{\partial \dot{q}_s} - \frac{\partial L}{\partial q_s} = 0, \quad (s = 1,...,n)$$

and only the knowledge of Lagrange's function is used $L = L\left(t, \dot{q}_j, q_j\right)$.

§2 Stability of the mathematical pendulum oscillations at the upper equilibrium position

We shall consider oscillations at the upper equilibrium position of the mathematical pendulum with length l and mass m, with the vibrating vertical point of support A. We shall define the coordinates of the centre of gravity of the pendulum by x, y, the pendulum angle from the vertical by φ and the deflection of the point A of the pendulum support in a vertical direction (figure 1) by $r(t) = a\sin\omega t$, where a is the amplitude and ω is the frequency of the vibrations.

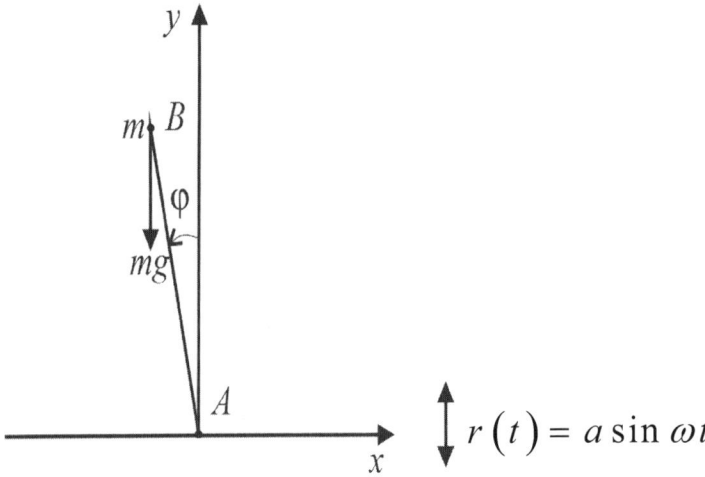

Figure 1

For the coordinates of the centre of gravity we find the following,

$x = -l\sin\varphi, \quad y = l\cos\varphi + r,$

and for the velocities

$\dot{x} = -l\cos\varphi \cdot \dot{\varphi}, \quad \dot{y} = -l\sin\varphi \cdot \dot{\varphi} + \dot{r}.$

The kinetic and potential energies are found to have the following forms

$$T = \frac{m}{2}(\dot{x}^2 + \dot{y}^2) = \frac{m}{2}(l^2\dot{\varphi}^2 - 2l\dot{\varphi}\sin\varphi \cdot \dot{r} + \dot{r}^2),$$

$$\Pi = mg(l\cos\varphi + r).$$

Then the Lagrange function is as follows,

$$L = T - \Pi = \frac{m}{2}\left((l\dot{\varphi} - \dot{r}\sin\varphi)^2 + \dot{r}^2\cos^2\varphi\right) - mg(l\cos\varphi + r) \quad (4)$$

and $\min_{\dot{\varphi}} L$ is found with the help of the equation

$$\frac{\partial L}{\partial \dot{\varphi}} = m(l\dot{\varphi} - \dot{r}\sin\varphi) \cdot l = 0$$

to give the following expression

$$\min_{\dot{\varphi}} L = \frac{m}{2}\dot{r}^2\cos^2\varphi - mgl\cos\varphi - mgr \quad (5)$$

As $r(t) = a\sin\omega t$ we get

$$\langle r \rangle = \langle a\sin\omega t \rangle = 0, \quad \langle \dot{r}^2 \rangle = \langle a^2\omega^2 \cos^2 \omega t \rangle = \frac{a^2\omega^2}{2}$$

where <X> is the time average of the function X.

Therefore the following is found,

$$\Pi_0(\varphi) = \left\langle \min_{\dot{\varphi}} L \right\rangle = mgl\cos\varphi - \frac{m}{2}\frac{a^2\omega^2}{2}\cos^2\varphi,$$

where $-\Pi_0(\varphi)$ is a dynamic analogue of potential energy.

The function $-\Pi_0(\varphi)$ has its minimum at the point $\varphi = 0$, if the following condition is satisfied

$$\frac{\partial^2 \Pi_0(\varphi)}{\partial \varphi^2} < 0 \quad \text{or} \quad mgl - \frac{m}{2}a^2\omega^2 < 0.$$

The stability criteria is reduced to the following inequality,

$$a^2\omega^2 > 2gl.$$
(6)

Relation (6) was discovered earlier in relevant scientific work. It is obvious that it was first discovered in some works [2, 3], and later repeated in the works of   N. Bogolyubov, P. Kapitsa, G. Stoker, K.Valeev, T. Stryzhak.

Abstract from the next volume

New Determinant Theory

$$e_m \cdot e_k = -e_k \cdot e_m \rightarrow e_m = 0$$

This brochure presents the Determinant theory with the use of hypercomplex Grassman figures. It enables to simplify the proof of many determinant features. It also states several ways of solutions of linear algebraic systems, most of which are connected with the Determinant theory.

The teaching experience in technical universities has demonstrated that the suggested way of presenting the Determinant theory is understood more easily.

The title of the brochure is explained by a desire to attract readers' attention to the well-known in the algebra sphere, but a new and more convenient for technical universities way to present the Determinant theory.

All your opinions, remarks as well as advice and recommendations will be taken into account to improve the next editions.

ibidem-Verlag

Melchiorstr. 15

D-70439 Stuttgart

info@ibidem-verlag.de

www.ibidem-verlag.de
www.ibidem.eu
www.edition-noema.de
www.autorenbetreuung.de